Crossing the Equal Sign

Crossing the Equal Sign

Marion D. Cohen

Plain View Press
P. O. 42255
Austin, TX 78704

plainviewpress.net
sbright1@austin.rr.com
1-512-440-7139

Some of these poems have appeared in: *The American Mathematical Monthly, The Mathematical Intelligencer, Bridges Conference Proceedings, Facets, The Journal of Humanistic Mathematics, Association for Women in Mathematics Newsletter, Elixir, Philadelphia Poets, Thirteen, Mainstay* (Well Spouse Association Newsletter), *Minetta, Meta Math!* (book by Gregory Chaitin), *Range of Motion* (disability anthology), *Mad Poets Review, Big City Lit, FOCUS (newsletter of the American Mathematical Society),* and *Conversing with Mystery: Poetry for Transition and Loss* (CD produced by The Hospice Recording Project).

Cover art:

Paul Klee, "Comedy", 1921.108, Oil transfer drawing, water-color and pencil on paper on cardboard, 30.5 x 45.4 cm, Tate Gallery, London

Contents

Points were blinking.
Lines were beckoning.
How was I to know?
What could I have done?

I heard some voices.
I had some time.
There was a tenderness.
There was a weeping.

How was I to know
the points would not point?
How was I to know
the lines would not line up?

I could think about those twittery lines
while brushing my teeth
or washing the floor.
So why do I want to just stand here
preferably sit here
maybe even curl in a crooked ball?

Why do I bend?
Why do I roll?
Why do I need to identify
my head with my knees?

I am not crying.
I am only thinking.
So why do I need
to be so small?

Yes, points were blinking.
Lines were flirting.
Spaces were trampolines.

I could have consulted the Math Reviews.
I could have leafed through a graph theory text.
I could, that is, have notified the authorities.

But I'm a do-it-yourself-er.
I'm a rugged individualist.
I'm a learner and a lover.
I'm a very foolish heart.

Someone wrote a book called *The Joy of Math*.
Maybe I'll write a book called *The Pathos of Math*.
For through the night I swivel
between intuition and calculation
between examples and counter-examples
between the problem itself and what it has led to.
I find special cases with no determining vertices.
I find special cases with only determining vertices.
I weave in and out.
I rock to and fro.
I am the wanderer
with a lemma in every port.

I feel so sorry for the insides of things.
I imagine them sweating and cramping.
I hear them trying to flex.

I know from Complex Analysis that sometimes outsides
    determine the insides.
And I think maybe the insides are tired of being determined.

And MOST things are inside.
Most things are encased.
I am afraid most things are alive.

I would like to go around rescuing all the insides.
I would like to dig into everything and pull the insides outside.
But there is not enough outside to go around.

If I can't rescue them, maybe I can put them out of their
    misery.
I know I can't shoot them.
But I can try to squash them.
Or I can go around injecting poison into them.
But what kind of poison works
for this form of life?

(The One-Dimensional Man: Some Questions)
Is his face along or across his body?
Is his mouth along or across his face?
Can his lips part? Can the corners turn?
What I mean is, can he smile?

One night I dream I'm dead.
Just standing there, dead.
I know I'm dead because my hands won't hold flowers.
I know I'm dead because all the sounds are a half-tone higher.
And the ground won't touch my feet.

The earth's surface is warped, not waiting for the horizon but
     shimmering right here, right beneath my eyes.
The sky's surface billows, has become its own planet, boiling
     up craters.
The lines, too, are wires and veins.
And whatever is not lines is growing lines.

Moreover, the world's only three-space is a soft grey whale
     squirming for escape.

I'm dead and no one's coming.
I'm dead and no one's calling.
I keep looking, keep shuffling.
I'm dead and I don't know what to do.

Once I've proved a theorem, I know it's true, that should be that.
But I keep thinking of things.

When your ankle is broken you're tempted to contract it
as though it had a muscle
as though it WERE a muscle.
"Tap me," says the bone. "Test me. Hurt me."

A theorem is muscles.
A proof is muscles.
Math is full of muscles
which, despite the pain, I'm tempted
to tap or flex.

There is a sibling rivalry
between this conjecture and its negation
and I, poor mother
throw up my hands.
"Anything, anything
"whatever you decide.
"Just please
"hurry up
"and make up your minds."

A mathematician just sits there.
There is empty paper in front of her.
And it stays empty.
A cat also just sits there with empty paper.
But the cat doesn't mind that.
The cat, in fact, gets right on top of the empty paper.
The cat believes his sitting will fill it up.

What drove me into math
was not Fermat's Last.
I preferred the factoring of the difference of two squares.
And Cantor's stretched-out one-dimensional lace.
Also, the center of a circle is inside the circle.

What drove me into math
was not the Mystery of the Unknown
but the mystery of the known.

Other early influences:
the point of light just happening to coincide with the only
    visible corner of our livingroom
those dark-red shapes when you close your eyes tight
and that spot, that nightmare
of many bloody colors.

I dream, these nights, of two straight lines.
Two inchers. Two pokers.
They have a common endpoint.
Or try to.

In some versions they nudge each other.
In other versions they peck at each other.
In still others they have
both endpoints in common.

I am getting to know these characters better.
But they ARE characters.
Elusive characters.
Like the woman next door they were friendly at first.

Still, they're beautiful characters, sweet and fun.
I like these characters, maybe need.
I am trying to know these characters best, but
this is as far as they want to be known.

The baby in the mirror smiled and cooed.
The baby outside the mirror giggled and goggled.
I wondered whether they knew each other.
I wondered whether they liked each other.
I hoped they wouldn't fall in love with each other.

And what did he think of the other me?
Which me did he like better?
Which me did he want to hold which him?
Did he think it perfectly reasonable, one mother being in two
  places?
Or did he know it all along, that every object consists of two
  equal and opposite parts?

We always say THE mirror.
We say THE mirror because we know they're all the same.
We know they're connected.
We know they're co-planar.
As one as windows on a single wall.

But the babies smiled and cooed.
It was as exciting as a trip to the zoo.
Except, one day I took him to the mirror and they DIDN'T
  smile and coo.
Didn't cry or tremble either.
Just puckered and smirked
sported that look
and turned the other way.

(Dream of the dragging of inertial frames)
I discover that objects can't move without pulling what's next
    to them.
So when a train starts up, so does part of the road.
Everything is like gum.
Everything is like muscles.
These muscles work very hard.

A train can't be moving and not the road.
I can't be moving and not Jeff.
I can't do range of motion on him
While he does range of stillness on me.

After awhile making a proof is like making a calculation.
There are certain things you automatically do.
You move with closed eyes, clenched eyes
unseeing eyes, no eyes.
You move, sometimes, with no brain.
After awhile crossing the implication sign is like crossing the
    equal sign.
After awhile a proof is collapsed to a point.

Today Devin catches me at it.
"Mommy, what're you trying to do?"
"Oh," I say, "well, these lines.
"I'm trying to fix these lines."
And then I explain triangles.
And then I explain transitive.
"Sometimes it can't be done," I tell him, "and other times it
     can.
"I'm trying to figure out when it can."
"I get it," he says. "I get it, Mommy."
And later he catches me at it again.
"THAT one WORKED, right?"

'Cause maybe, if that one worked, we can go play Parchesi.
Or cards. Or ice cream.
Or at least Mommy won't
keep staring at those lines.

(Teaching Sturm-Liouville)
I multiply the mu equation by nu and the nu equation by mu.
But not enough goes away.
So I multiply both equations by zero
and then subtract, just in case.

It's all zero.
Points are zero, lines are zero, the set containing zero is not
    one but zero.
There is nothing to learn, nothing to teach.
There's been a terrible mistake.

(Dream of Math Research on Anti-Equivalence Relations)
There are two sets of four objects each.
Within each set each of the objects
has nothing to do with any of the other objects.
Furthermore, there are two objects, one in each set
which have nothing to do with each other.
What conclusions can you draw?

Today we were alone in that room again
I and this man
Who was once my world
Later my burden
Lately my enemy
And still, somehow, my husband.

Alone in this hospital room.
Alone these soft evening hours.
Each of us has something that is hurting
And that we need to get rid of.
For him, life.
For me, the problem of characterizing comparability relations.

He lay there, eyes open but unseeing
Mouth yawning but un-tired
And possessing no symmetry about any line.
I sat with him, healthy
But equally stopped, equally unable.

Yes, there we were again, working side by side
Kneeding and grating our separate wounds
Wishing against wish that time could heal them.

Every once in a while his eyebrows twitched
And every once in a while I wrote something down.

Never judge a romance by its ending.
The end of romance is NEVER any good.
Endings are chronic, endings are progressive, endings are full
of excacerbations.
Endings are full of what has happened.

Endings are full of very sad givens.
Endings are full of nothing to prove.

Every 4:00 A.M. Substance is the first of six cats to arise and
    go rummaging through the wastebasket.
He needs something, anything.
It doesn't have to be food.
As long as he can chase it around.
As long as it scratches along the floor.

Devin says he's frightened.
Frightened of everything.
But I say Substance is looking for something to prove.

So give him definitions.
Notation.
Axioms.
Yes, Substance needs some axioms
to scratch along the floor.

(Nightmare that we write in a language in which every shape
is a letter, so --)
we can't doodle
can't cross out
can't make any stray mark.

It's like trying to throw out a trash can
Or cry wolf
Or leave a phone message
When the machine is full.

One day the baby discovered that his index finger was a
   unit vector.
Lamp, mirror, window, sky, all became destinations.
"Why don't you POINT to the ceiling?" I lingoed. "Why don't
   you POINT to the floor?"
All just purely for the pleasure of studying the rotation of that
   chubby little Cartesian frame.
But then, quite suddenly, I was on the wrong end of the arrow.

To whom was he pointing me out?
And did he mean to inform me that he wasn't what he seemed?
How scary a word, how chilling a deed
"you" actually is.

Another mathematician I know does nature photography.
She says she likes to pretend she doesn't know what's in front
    and what's behind
or what's a reflection and what isn't
or what are the objects and what are the spaces.
She says she likes to look at things
as though she doesn't know what they are.

(The Ballade of the Theory-Creator -- "There are the theory-
    creators, and the problem-solvers"…
    … Halmos)
"Problems worthy of attack
"prove their worth by fighting back."

And being obnoxious, mean, and nervy.
But I've something just as worthy
just as wondrous and exciting
proving worth by more than fighting
staying content to work WITH me
in peace and love and harmony.
It doesn't hide, doesn't cower
doesn't try to snag the power.
We pass the wee hours side by side
with truth and beauty as our guide

stay at each other's beck and call.
I don't have to attack at all.

(First try at Zen meditation)
We let out our bottoms
hold in our tops.

We count our breaths
mod ten.

And the flow of the room is counter-clockwise
like the flow of most mathematical surfaces.
Is anything ever clockwise?
Only the clock, the lone clock.

Only the clock sails counter-counter-clockwise.
Provided it's not digital, it persists
insists
and resists
the flow of the room.

Who needs Agatha Christie, today
when it is not yet known
whether every set can be ordered
whether that every set can be ordered is equivalent to the
    Axiom of Choice?

And who needs Sharon Olds, who needs Anne Sexton
who needs my own poor worn-out words
when I simply would rather
simply keep turning
keep being turned towards
this?

And who needs thrift-shopping, this week
when there are all these MATHEMATICAL objects?
And you can put subscripts on them and make them even
    more like objects?
It doesn't have to be geometry.
It can not take up space and still be pretty.

continued

And who needs love
when it's all right here?
And there are, and were, those who PUT it here
a whole bibliography of Tortured Souls
people who care whether or not every set can be ordered
or who don't care
must only know?
Yes, who, alas, needs love?
Who, right now, needs a man or a child?
when I can, and do
lingo
and whisper sweet nothings
to this
about this?

(The Stuttering Mathematician)
In order to get to the first word
you have to get to the first half-word
and in order to get to the first half-word
you have to get to the first fourth-word.
So how can ANYBODY do it?
get smaller than syllables?
smaller than letters?
How does ANYONE begin?

One morning I wake up counting.
I realize I'm counting and I just keep it up.
There's nothing to count.
Nothing to own.
Nothing to try to win.
I'm practicing counting
preaching counting
maybe even learning to count.

In the dream I had not been counting.
I had only been wandering.
The horizon had been approaching
and the sky had been flat.
Also, that sky
might have been folded.

There's no reason to count footsteps.
No reason to count years.
Maybe no reason to count numbers.
To fall asleep you count sheep.
What do you count to fall awake?

I am no workaholic.
But I'm collecting points and lines.

Not like stamps.
No, I wouldn't trade them.
I simply have to have them.
I need a group portrait
all of them smiling.

I have to have a hand
with these beauties as fingers.
I have to hold a vase
with these cuties as flowers.

I should contact a colleague.
I should go online.
But -- don't you see?
I have to do this alone.

I am based in reality.
But God created these lambies
set them out to green-pasture
and maketh me
to lie down.

What is this business of things existing?
What is this business of people existing?
What is this business of math existing?

When I get that far gone I imagine a piece of paper with math
    written on it.
I imagine cutting out the math
cutting around all the numbers and symbols.
I imagine the cut-out math and I imagine the stencil.
The paper is very white.
The math is also white.
Maybe I even imagine cutting out the math without it having
    been written.

(A rhyming poem, written after fleas)

I used, at least, to not question specks.
X was the spot; X was just X.
A speck could be unexplained, if it must.
A speck could pass for dirt and dust.

Points could blink. Points could jump.
And in the night points could go bump.
But a speck was an axiom. A speck was a mother.
It could not disappear from one place and appear in another.
And a speck did not move like a shrunken train.
And a speck did not contain a brain.

This week, though, I've gained some knowing.
This week specks have gained some going.
There's been some movement in my picture
been some compounds in my mixture.

I've seen tiny shadows on my drapes
and noticed points with different shapes.
I have learned to distinguish black from brown
learned to distinguish flat from round.

I know all the nicks in our kitchen floor.
I know which ones were here before.
I have memorized each and every spot.
I know which belong and which do not.

I have to admit it: These lines are abusing me.
Or someone or something is wielding these lines.

Yes, the dentist has placed extra teeth in my mouth
and the night has placed extra lines in the plane.

When nails grow too long you cut them.
When a man gets violent you leave him.
But when lines take aim
and form a fence
or when points zing far and away
what
do you do?

I am no addict.
I am no psychotic.
I am no Woman Obsessed.

I am no Plath.
But auxiliary lines keep sprouting and thickening.
I am no Sexton.
But this black art will be the death of me.
I am no Poe.
But wouldn't it be something
if math turned out to be bad?

I am no schizo but these lines are beginning to schiz.
I am no werewolf but each space might contain a moon.
I am no vampire but this is a matter of blood.

I am no cafe-genius but afternoon has turned midnight.
I am no suffering-artist but this is science.

I am not O-D-ing but somebody
please
come quick.

It's a kind of transitive law when
in a house of growing children
two people who pet the same cat are petting each other.
Especially if one of them is holding the cat.
Especially if both of them are holding the cat.
And if Devin gets under the blanket with Mirage
and lets only their heads stick out
and smiles up in that way
if the pug of Devin's nose is close to that spot between
    Mirage's ears
and I grab hold of it all
and kiss it all. . .

well, Devin also knows
and Mirage also knows
that something is necessary
something is sufficient
and something else is scared.

These lines didn't ASK to be here.
They simply NOTICE that they are here.
And so they sizzle.
And so they sway.
Like a bag of spiders.
Like a litter of mice.
They are surprised.
Maybe scared.
If they had eyes, those eyes would widen.
Those eyes would have no brows.
They might also have no mouths.
Or their eyes and mouths might be the same.

(Dream that the binary function, distance between,
    is not symmetric)
There are human condition metaphors --
It takes longer to return than it does to leave.
I move closer to someone who moves further from me. --
But mostly, I stand at the top of a slide and look down.
Then I stand at the bottom of the slide and look up.
And I take a stick and twirl it, watch it changing size.

I dream the distance between a point and itself is not zero.
Each point shivers.
Each point is exiled
from its small country.
There are preferred directions.
There is a great wind.
A general current
has begun.

This is not the dragging of inertial frames.
This is the RACING of inertial frames.
Space is proven not to exist.
Everything looks for a place to go.

(Dream about trying to unload a gun)
Mechanics never was my forte.
I'm a THEORETICAL mathematician.
I try this compartment, that, but nothing clicks open, nothing
    even shifts.
Finally I hear a muffled sound.
But they drop out slowly, tantalizingly, and not far enough
    away.
And maybe I left one in by mistake.
I keep shaking, jiggling.
They're like salt from the shaker or glass on the floor from a
    broken bowl.
I'm scared enough to keep jiggling, to stay in that little room.
And I stay scared until...
well, until nothing. I'm still scared.

(First Day at the Beach)
After the heat of the day I get past the breakers to where I can
    graze
where I become aware that we are all of us in the lap of
    Mother Earth
where and when I can believe in Earth as a mother
playing with us children, playing with us kindly.

But then I catch sight of the horizon
its slight frown.
And I see that I am looking down at it
not across.

I see that Earth is a CONVEX mother.
We are on, not in, her belly.
It is not us she is in labor with.

Not that we'll fall off.
Only that there is that peripheral vision.
There are those two wings
and they bend, then sway.

And perhaps we are the balancers rather than the balanced.
Perhaps we are tightrope walkers
bearing the x-axis.

("An die Mathematik" -- after the Schubert Lied, "An die Musik"
("to Music"))
I have known grey hours.
I have known wild circles.
And you, too, have betrayed me.
You, too, have erased me.
You, too, have been worthy of attack.

But you add texture to existence.
You add mind to meditation.
And you illustrate my diary
with your better greys and wilds.

And I thank you for this, I thank you
whether or not I'm welcome.

(listening to Händel, #4 in D)
They're dual ocean waves
a war between sine and cosine.
And I know that the instruments are together
(They wouldn't allow them on the recording if they weren't.).
But they don't SEEM together.

There's a rushing, or a dragging.
I keep hearing distances.
The distance between the violin and piano.
The distance between the piano and the microphone.
The distance between the microphone and me.
And the distance between my two ears.

And if I try to contract the distances to points
it all still goes sloshing around.
I want to pause, gesture -- "Let's start ALL OVER"
but wind up throwing out my arms in resignation.
"Well, Einstein SAID there's no such thing as simultaneity."

But when I PLAY the Händel, the distances disappear.
Everything is as simultaneous as it pleases.
Time is not relative but quite absolute.
I'm the origin, or middle-C's the origin, we're all the origin.
Everything is identified
with the same point.

continued

Especially that last chord.
"Okay, now, GO" and it all pulls together
same time, same space.
It all comes up, one big event
perfect, precise, not slipping and sliding at all.

(Portrait of the Mathematician as a Young Woman)
Addition is commutative.
Multiplication is commutative.
How come exponentiation isn't commutative?

Or: Exponentiation isn't commutative.
How come addition and multiplication are commutative?

What a quest, for a sixteen-year-old kid.
What a journey, for a seventeen-year-old adult.
What kind of hormones were those?
What kind of raging was that?

The sky was wrinkled and old.
I was smooth and young.
But that starless twilight trees were dissolving into dark
    powder
Flagstones were trap doors into the earth
My parents' backyard spread out like a sanitarium garden
And I cupped the face of that sky
In my vibrating palms.

I held that face as though it were a magnifying glass.
Then I stroked it with one flat hand.
Finally I waved both arms over it again and again.
There was nothing else to do.

Yes, distance brought us together
That 7:30 P.M.
Distance and the things
We each didn't know.

(Mozart)
Sometimes I don't want to accent the first beat.
I want to accent ALL beats.
I want to play every note as though it were the first.

Every MATH-thing should also be accented.
Equations, expressions, single Greek letters.
Well-formed formulae, meaningless strings of equal signs.
High-powered theorems, incorrect ideas with no applications.

Every truth is beauty.
Every un-truth is beauty.
I love all, I love each.
Should write Theorem before, QED after
a G-clef before, a double-bar after
each.

Händel again, this time singing.
And problems again, simultaneity problems.
There's a rest, a dotted-whole rest.
When it's over, will the pitch come together with the word?
And if so, when?

I'm tempted to make it too soon.
And I'm tempted to make it too late.
As for just right, that's a moving target.

I count
keep counting
counting to the true now
the now to end all nows.
It's a rest, a long one, but I'm not resting.
I'm hovering.
The correct now is a point
smaller than the head of a pin
on which not even one angel
can dance.

Is math hungry?
The way I always say a baby's hungry?
And the way Anne Sexton says "O my hunger!"?

Oh, I know I once wrote "Math is a cat; feed it" but does that
    mean math is hungry?

Well, SOMETHING is hungry.
And math, maybe math
is thirsty? Aghast?

Or is math the food?
The plates? The spoon?
Maybe math is the stomach?
What should be fed? And how?

(The latest math dream)
I love one over n minus one over n-plus-one equals one over
    n times n-plus-one
along with its proof.
But tonight I revere it so
that I write it, for the case n = 3
on the birth blanket just before birthing.
It is so urgent that the pen is scratchy
and the blanket loosely woven.

I make a ceremony of it
a celebration.
I draw that very long fraction line
slowly.
I begin the line in its middle, smoothing outward
tenderly
like the laying on of hands.
I make that fraction line a plane.

Once an editor wrote "sorry, can't help you".
I don't need help.
I am only a dreamer of math dreams
a writer of math poems.
And I changed my mind, I DON'T love one over n minus one
    over n-plus-one equals one over n times n-plus-one.

continued

I love the difference-quotient, the product rule, formulae with
    little epsilons.
Fractions are too elementary to love.

But couldn't somebody please just WATCH me have this
    baby?
LET me have this baby?
At least help me press down this cloth.

Allow me just one more
just a single unit more
let me try the formula
for just one other n.

(Dream of the woman with a four-dimensional lover)
He literally drops in on her every once in a while.
Lonesome as a ceiling, lost as the sky, he sags, first over, then
    onto, her head.
He literally rolls into her life.
An over-sized snowball wanting more and more, he attempts
    to gather her.
He literally doesn't travel in her circles.
Shining on her like the sun, he is aware that only his rays can
    touch her.
Spinning like the moon, he carefully selects which face to
    present.
There is a place down the cellar where she goes whenever
    she really needs him.
Wonder-Woman style, she flees to the grey windowless room
    and stands next to the boiler.
Then she raises her head and spreads her arms.

But she doesn't have to spread them far.
And she doesn't have to spin.
And if there is a party down that cellar, and if guests have
    seeped into that room
she just waits for a turn of the conversation and literally slips
    away.

continued

His relatives have something to do with it.
They adore her.
Although they speak Italian and she barely at all, there's a
    rapport.
His father, in particular, slaps her on the back.
And his grandfather brings out the old family album.
His uncle will escort her on a tour of the house.
Before mounting the stairs, she turns once more to glance
    back at her lover.
He smiles and nods; he'll be waiting for her at the foot of the
    stairs.
She smiles back but hesitates.

For well she knows the dangers. The dangers of waiting
the dangers of stairs
of time
of space
and of lines, no matter how short.

I stand up there and dance
the dance of the unsteady.
I sway, I swirl.
I spin a semi-circle
afraid of making a mistake and spinning a full circle.
Yes, afraid of integrating from zero to two pi instead of just pi.

Oh, teaching is a dance.
A fast dance.
The dance danced by the vibrating string.
The dance danced by any Derichlet f over all its sine's and
    cosine's.
Yes, teaching is all dances.
It doesn't miss a step.

I stand up there and wobble
like dark in the hall.
I stand up there and totter
like a chandelier in a storm.
I'm the fancy dancer, the frantic dancer
the answer dancer, the cancer dancer.

On, Dancer! On, Pancer! On, Prancer! On, Chancer!

Yes, teaching is some dance.
It goes on and on.

continued

And I stand up there and run.
The curve, not of pursuit, but of flight.
And you know The Loneliness of the Long Distance Runner?
Well, this is the SHORT distance runner.

Yes, teaching is a sport.
A competitive sport.
I throw, I catch, I aim, I dive.
But mostly, I run.
I run the way sine n-nought x runs through the Fourier series
knocking down everything in its path, searching for itself and
    then running on anyway.

I run through the theorem, run through the proof.
Run through the lies, run through the truth.
I run through the air with the greatest of ease.
Just like the rain smashes down through the trees.

(teaching curve-sketching)
Local min's make smiles; how curious, that low points mean
　　happy.
Also, smiles and frowns must alternate.
And between a smile and a frown
must be at least one smirk.

Omigod, that local max looks an awful lot like a breast with its
    nipple.
And the graph of $x + 1/x$ is big-time phallic.

And THEY're pretending not to notice.
So good in their seats, such straight faces, not a smirk on
    them.

And suppose they decide to stop pretending?
Suppose they decide to stop sitting good and start sitting bad?
How can I fix that local max?
In what coordinate system is the graph of $x + 1/x$ not phallic?
In what coordinate system is nothing male or female?
In what coordinate system are all curves straight lines?

(A dream about police and "We guess a solution of the
    form…")
"What makes YOU decide to guess a solution of the form X
    times Y?
"No one said anything to YOU about X times Y.
"What is it you've done that makes you say X times Y?

"And so hastily.
"So efficiently.

"I think we'd better have a look around and see what's up.

"Yes, we have a search warrant.
"No, you haven't the right to remain silent."

I have said that math can feel nostalgic.
But these nagging lines
and the shapes they try to enclose
are not the dancing Klee figures
under my clenched eyelids
nor that toddler nightmare
of the spot of many colors.
Rather, these hellkites
are in the PRESENT tense.
They're twigs
or stems
with flowers at neither end.

The hardest job in the world is not being a mother.
The hardest job in the world is characterizing comparability.

Oh, the first round of determination works fine.
It sleeps us well.
And in the morning
are equivalence classes.

But in the evening
the points run amuck.
The lines go astray.
They all fly into
another back yard.

To whose home should I call them?
What do I have for dinner?

I save unused lemmas.
No matter how silly.
Just like I saved my old diary.
I was fourteen and outgrew it.
But I would not divorce it.
I would not banish it.
I certainly would not kill it.
Instead I put a sign on it.
"Never throw this away.
"Never throw this away."

(The Successful Stutter)
One day I suddenly just-couldn't ask whether this train
    stopped at Willow Grove.
But I stayed in the middle of the just-couldn't.
I watched the guy watch the just-couldn't.
And the just-couldn't lasted awhile.
And he and I just-listened
just-waited.

I wasn't exactly relishing.
I was even turning away.
But he, I, and the word
formed a little triangle.
The way the first three fingers of a hand
can bunch and then huddle.
The kind of triangle formed by refugees in a storm.
Namely, not three sides meeting in three points
but three lines emanating from one point.
In other words, not a triangle at all
but a three-pointed star.

Today I teach parabolas.
Soon I'll teach loops and scallops.
And sin one-over-x with its big and little fusses.

But I also remember straight lines.
I mean straight lines without axes.
Straight lines without slope.
Straight lines that aren't tangents.

I remember trying to prove Euclid's Fifth Postulate
and coming close
finding smaller and smaller triangles
little lights passing through a diamond
eensy weensy triangles
climbing up the wall.

Oh, calculus says that curves are okay
and they are
I agree.
But I haven't forgotten straight lines.
I keep coming back to straight lines.
I'm still not finished
with straight lines.

Someone just told me that Mozart was thirty-four when he died.
I had thought he was thirty-seven.
So now I lie awake, then sit and wander awake
counting again
subtracting again
grieving for those three years.

(Dream of Two Vacation Licenses)
(One) If, around noon
I decide I don't like the day
I'm allowed to go back and dream
that it's 7:00 A.M. again.
I can do that as often as I choose
so that, by induction
I eventually get a noon I like
and not the set containing the set containing
that noon.

(Two) I can begin the day by covering myself
first with any color I like
then any other color I like.
Then, when I've done all the colors
I can apply the last coat, black.
Then, around noon, I can scratch
with the point of a gentle scissors
any design I want

making pretty flowers
hearts and smiles
parametric curves
rainbow equations
as pretty and as true
as I damn please.

We begin with an equation
and then we do everything we need to solve it.
And not only is the equation equivalent to the solution.
It's equivalent to every equation along the way.

The important things are the equation at the beginning and the
    solution at the end.
The unimportant things are the equations in the middle.

They're little lemmas.
They're there when we need them.
And when we don't need them
they stay out of the way.

One long February my right leg tried to die.
It tried to lie low, to stop wriggling its toes
to simply fall against the inside walls of that cast

to stop pounding inside its epsilon-neighborhood
to stop trying to get epsilon on all sides
to settle for two epsilon on one side, zero on the others

better to forget, to sink into the bottom, zero on ALL sides.
It tried not to toss and turn all night.
It tried not to keep making decisions as to how to distribute the
    epsilon.

It tried to die, and to stay dead
for the next four weeks.

I don't invent math.
I don't discover math.
I only PLAY math.

Like I play the piano.
Playing Partial Differential Equations is like playing Lizst.
Playing Complex Analysis is like playing Mozart.
Playing Abstract Algebra is like playing Vivaldi.

Or perhaps I PRACTICE math.
I practice and practice until I'm ready to preach.
Then I preach and preach until I'm ready to perish.

Is there phrasing?
Is there dynamics?
No, but there's fingering.
And there's running out of fingers.
And there's shaking out my hands
and starting anew.

Finally the cast was cast AWAY.
Out twittered the epsilons like little birds
off they poofed like soap bubblets
and in zoomed aleph-null from all sides.
A circle, a sphere, with radius aleph-null.
There was aleph-null forever.

Aleph-one came also.
Beyond space, beyond the ruler, beyond the searchlight
but, for the occasion, it bent in anyway.
And then aleph-two, and aleph-three.
They all came in to kiss, to give blessings.
And aleph-twelve, that thirteenth fairy, was heartily invited.
All the alephs came and stayed; I felt the winds of their
    coming.

The foot sagged like a belly after childbirth
or the lead shield they use when they take X-rays
or a pancake when you've forgotten to put in the baking soda.
The foot was not quite back.
It had been better at dying than it had thought.

It couldn't yet do much with the alephs.
It would have to use the walker.
But space is not homogenious.
Each point has different alephs.
I couldn't wait to take the walker and try
the alephs of the corridor, the elevator, the X-ray room.
I wanted to try them alone
taking my space and my time.

The main trouble with adult life is things like purses.
And when we walk into a room we have to remember what we
    went in there for.
Also, when we walk out of that room we have to think what we
    can take out, while we're at it
and what room to go into next.
And stairs. We're always parking things on stairs.
That way, on our next trip up we can stoop down for
    something to take with us (while we're at it).
When we step into the shower we take off our watches.
As we drop off to sleep we memorize where our purses are.
Yes, once we become adults our bodies aren't constant.
They're plus or minus x, where x equals purse, watch, ring,
    Kotex, diaphragm, wheelchair, colostomy bag.
We die humanoids. We die keeping track.
Our bodies ourselves
our purses, our pills.
We die checking, we die memorizing.
We die pulling on, we die flinging off.
We die knowing we haven't
a thing to wear.

(Imagining Buddhist Judgement Day)
The enlightened universe lolls around.
It is big and fat and one and only.
It's a little homesick.
It misses parties.
It misses babies.
It misses sweaters.
And it remembers math.

Points used to be blinking.
Lines used to be trembling.
Now it's all homogenious.
Also, one of its corners used to be me.
And another of its corners was my true love.
And there was time.
There was the first this and the last that.
Now it's all middle.

The universe is bored.
Isn't there another universe around somewhere?
Is this all there is?

(How I Know This Isn't a Dream)
I believe what I'm told.
And I believe what I read.

And besides, this table feels pretty solid.
When I knock-on-wood it's there.

True, points are blinking.
And lines are squirming.
But planes hold steady.
Planes hold flat and firm.

And my hands fold nicely.
My fingers wriggle well.
And here I am staring
at the center of my palm.

Maybe leaves are ghostly.
But trunks are not.
Also, I widen my eyes
and I don't wake up.

(Dream of Two Women Sharing a Wheelchair)
Are they friends, lovers?
Are they Siamese twins?
Or merely too poor to afford separate seats?
Are they side-by-side or front to back?
And is one a lot taller?
Or one a lot wider?
Do they hold hands?
Do they rub feet?
And IS it just two?
Is there some small fraction
of another woman?
And if so, where?
Between them? Behind them?
Is she around them?
And how many times?

A mathematician should never watch action films.
She has already swum through iron, run without roads, flown
    without sky
has already known too many directions
has already been reduced to a point.

She has had enough of thinking hard
enough of hoping that thinking will save her.

(Extra objects)
They come in the same boxes as the objects we ordered.
They fell in by mistake.
They're not styrofoam.
Sometimes they're curved at one end and straight at the other.
They are not always little but they might be bugs.
They are not always egg-shaped but they might be eggs.
They might be poison, they might be bombs.
Or corners from the fifth dimension.

We don't want them.
But we're afraid to throw them away.

John Conway looks for things to prove.
Things to prove keep looking for me.
These things are a burden.
They render me lost.
They render me locked.
They render me away.

Lectures to prepare.
Onions to chop.
These are things
but not to prove.

I have things to prove, things to prove.
Like "caps for sale, caps for sale".
Could somebody prove my things?
Could somebody prove even
a red
thing?

(Portrait of the Mathematician as a Younger Woman)
Mr. Magic 4 isn't as interesting as Mr. Magic 9.
But I thought it was.
It was certainly more interesting than geography or American
    History.
So I sat at my little desk pouring out 12, 16, 20, 24...
3, 7, 2, 6...
I soon saw the pattern but I kept adding the digits.
Then I tried Mr. Magic 5.

There were no points.
There were no lines.
But something was beginning
to throb.

Remembering unused lemmas is like remembering childhood.
Maybe even deeper.
Maybe even thicker.
Maybe like remembering
something that wasn't there.

Math memories are different from other memories.
They're less clear.
And I ask more questions.

What did I write in the margins?
And were the margins big enough?
Was there a point blinking?
And was it on the page?

Yes, I ask more questions
because I want more answers.
And I want more answers
because there are more answers to want.

When you need more than you prove, it's a nightmare.
But when you prove more than you need, it's unnerving.
You don't need more
than one revelation.
You don't need more
than one excuse.

If you ask, "Why math?"
I'll say "same as science fiction".
Same fuss. Same fury.
Same stretching over the universe.

And not only infinity.
But each and every count.
Especially the single digits.
Each, separate, a pearl.
Each, separate, a face.
A rose, a bud. An insect, a cell.
Also, each a question mark
in some language.

(Math on the bus)
The only available seat is half a seat
because the woman on my right needs the other half.
Actually, the man on my left also needs the other half.
Still, using the calc text as a table, I manage to get settled
    working
And the woman on my right shouts out a monologue for all to
    hear:

"I FINISHED WITH THAT CALCULUS A LONG TIME AGO. I'M NOT DOIN'
NO MORE O' THAT CALCULUS. MY DAUGHTER WENT TO COLLEGE
AND I HAD TO HELP HER WITH THAT CALCULUS. I'M NOT DOIN' NO
MORE SINES. I'M NOT DOIN' NO MORE COSINES. NO MORE O' THAT
CALCULUS. NO SIR, NO WAY."

A block before my stop I put my papers away and look up.
The woman smiles at me and asks, "Did you give up?"
I smile back. "As a matter of fact, I did."

The idea is not to solve.
The idea is not to learn.
To understand the cosmos
is not the idea.

The idea is to think.
The idea is to do.
To rub a stone.
To pet a dustball.
To love a patch of space
even if it's empty.

And to love that for which
there is no space.

When what worked for years suddenly plays
when what played for years suddenly dies
when insides are not all that need to be rescued

when the ports in which we deposited our lemmas
suddenly become portals

when a theorem has neither application nor implication

when the easiest lemma to prove
the one you were not even proving
the one you were saving for last
when that lemma isn't true

a mathematician doesn't give up.
A mathematician insists on insisting
"SOMEthing is going on."

(On becoming a grandmother rather than a mother)
Why is my belly off to one side?
Why can't my belly be in my lap?
Why can't I get behind my belly?
What is in the way?

What is this angle?
What are these coordinates?
What kind of closed system is this?

How did my belly become so surreal?
Such a passing through, outside, numb belly?
How did my belly become such a shape?

Yes, what is my belly doing with my belly?
Is my belly going to run off with my belly?
Or will it bring it back and hand it to me?
Am I the surrogate mother or the adoptive mother?
Am I giving birth to twins or not at all?
Who is giving the gift?
To whom?

Theorem or counter-example, it does not matter.
No, it does not matter
whether or not squiggle is admissible if and only if there exists
    no odd determining chain.
It does not matter whether.
It matters only how.

If I know the end of a movie, I still have to watch it.
I have to see the middle.
I have to see it through.

(Math Research, Age 4)
One day I set out through the Queen Anne's lace.
There was no end to what I set out on.

After awhile the grass got sharp.
And the white field became like outer space.
Later still it rained
slobbering the sharp wet grass against my legs.
Destination faded into rumor
and so did home.
I was stranded in the middle.

But I was not lost.
The way back was the straight line behind me.
I was only STUCK
on the straight line before me.
I knew the way back.
It was the will I couldn't find.

All spring and summer the Sheridan Avenue gang would sit in
    the driveway next door as though it were a beach.
We'd rummage through thousands of pebbles, all of them
    typical
except for the pink one.
The pink one was rough, with just enough grey sparkle.
Sometimes we'd put it in a pocket and carry it around all day
and in the evening we'd throw it back.
When winter came, snow covered the driveway.
But in April the stones had not disappeared.
The pink one was the first to have not disappeared.
It had been like the same drop of water evaporating and
    raining back.
Like a theorem, it was still true
no matter how big the numbers got.

(Childhood evil)
We were having tuna fish for supper.
I brought Mommy the can.
Then I rested my chin on the tabletop.
"Oh goody," I said
and Mommy turned around
with the same smile.

But then I started sticking my fingers down my throat.
A gurgle, a gag, a wad of brown acid.
"Stop that," Mommy said, and I obeyed.
But then I started doing it again.

And I kept on doing it.
I kept on touching what I couldn't see.
I kept on creating what I couldn't destroy.
I kept on destroying what I couldn't create.

For a long time I knew the moon very well.
I knew the moon better than anyone ever had.
I could have written a textbook on the moon.

I knew, for example, that ten nights each month
the moon would visit my room.
It would survey the geometry
and fall like a transversal.
And I knew that it would stay longer
knew, each night, exactly how long.
I knew where, in my room, the moon would go
and I knew when.
What hour it would be a broken parallelogram over in that
    lower corner
what hour it would slither onto my bed
what hour it would have arrived on my pillow.
And I knew when it would wander
what hour I would have to climb out of bed
how long, each night, I would have to stay awake.

I knew the moon as well
as I knew my life.
I knew the moon too well.
I knew what it would do.

Before any of that I stared down a hole.
It was brimming with ink.
A can of night.

Maybe death is a can
and you fall into it
and someone slams on the lid.
Maybe millions of years go by
then someone opens the can
someone with a fork.
Maybe death is being canned
or gradually, gradually
becoming un-fresh.

Not ON 22nd Street
but IN 22nd Street
I saw a flattened cat.
It was bigger than a non-flattened cat
and skewed.
A parallelogram of a cat
with two dots for eyes
two circles for nostrils
a racoon of a tail
and less than four paws.

Picasso's cat.
"Still Life with Death".
A topological projection of a cat
with only a smattering of red, a smidgeon of wet.

I saw that arrangement
the way one sees
an eagle in the sky.
And I thought, "Not Tygra. Tygra's been missing only a couple
    of days."

But now she's missing a couple of weeks.
And the flattened cat is also gone.
So now I have two pictures, two cats.
They're side by side.
They have the same colors, same corners.
There is an isomorphism between them.
They are not super-imposed.

Maybe dying is like being arrested.
And they don't give you even a second of orientation.
Maybe they nail your feet, right off, onto the board.
Or they impale you on a skewer
half a second before
half a second after
two of the others.

On the other hand, maybe dying is like going to the doctor.
Maybe there's a receptionist and a waiting room.
And maybe there's also the assistant.
And she explains the procedure, shows you diagrams, holds
    your hand.
Yes, maybe she tells you first
what they're going to do.

I think I've solved the problem.
But I still lie in a ball.
What I wait for is time.
It has to pass the two-day test.
And it might not.

Well, Kerin didn't.
Kerin was merely born by me.
She did not get
to stay by me.

So I lie here and curl.
I lie here and wait.
Not for a phone call.
Not for an ambulance call.
Only for my own mind
to think too much.

Right now I'm thinking about black lace.
Black lace skirts, black lace scarves, two knee-high cartons of
    black lace knee-highs three pair for a dollar at Jo-Mar's.

Not black lace like a spider's web.
Not black lace like the top branches of a tree
superimposed upon the top branches of another tree.
No, I mean flower lace, birdie lace
miniscule dots and circles lace.

Not sexy lace.
Nor girdles and garters.
Not long legs covered with black roses.
I mean CUTE black lace. Sweet black lace. Feminist black
    lace.
Anyway, black lace.

As though it would be okay
if everything were all inked in
and all big objects were the same flower
and all small objects were the same bud
and the parts in between were netted

if space alternated like that between dark and dim
as though that would be okay.

(Another Thing I Like about Math)
It doesn't get written in cursive.
Each little bug is free to crawl.
Each little stone is free to shift.
Each is itself.
A fresh start.
And writing is as simple
as before third grade
before every plunge down
became such a commitment.

Remember Eureka?
Well, what's Greek for "I lost it"?
Did Archimedes ever hit the streets with THAT word?
Did he shriek it? Bellow it? Did the cries diverge?
Or did he sob and tremble it?
Bay it at the moon?

And his bathrobe -- did he pull and tear at it
as though it were a straitjacket?
Did he weep into it?
Clutch a corner and crumple?

And then, eventually, did he pick up again
into the next street
under the next sky
muttering that one Greek word?

A theorem is not a child.
No, a theorem is not a child.

Should I return the lemmas
to their various ports?
Should I then become the wanderer again?
Or should I just go home?

It's better to have loved and lost
than not to have loved at all.

And I wouldn't have had the Envelope Lemma.
I wouldn't have had the Two-Triangle Lemma.

I wouldn't have walked those misty streets.
I wouldn't have met those crusty old souls.

I have a new idea
so I get to make new lines.
Fresh clear runners
like tadpoles in the brook.
At first there are only six.
Then there will be twelve.
Soon the paper will resemble my jar of straight pins
some of which are bent pins.

In other words, these lines
will soon be dangerous.
They will be needles.
They will be knives.
Knives with no handles
with blades at both ends.

Eureka!
Pretty Eureka!
Pretty Eureka with sugar on top!
I have read the signs.
I have broken the code.

I collected my lemmas from every port and brought them on
    board.
I brought them to my country.
I see the scene, I see the act.
I have not solved the cosmos but I have solved this house.
Most of infinity is still unsolved but I have this picture.
I have this brain.

You can draw pictures without analytic geometry.
Straight lines don't need ax + b.
Circles don't need x-square plus y-square.
Loops and scallops don't need polar coordinates.

But they wouldn't be as pretty.
They're prettier with axes running through them and equations
    running along them.

Spirals aren't as pretty without theta.
Four-leaf clovers aren't as pretty without sin-square theta.
Beauty isn't as pretty without truth.

Points are not blinking.
Lines are not waving.
No, this new problem is not about space.
Not yet.

ABOUT THE AUTHOR

Marion Deutsche Cohen received her math Ph.D. from Wesleyan University in Middletown, Connecticut, and is currently on the adjunct faculty of the University of Pennsylvania. She first became passionate about math when she took her first algebra course, or perhaps before that. Besides the usual concerns that adolescents write about in their diaries, she expressed her feelings about math. Some of her other books are "Dirty Details: The Days and Nights of a Well Spouse" (Temple University Press), "Epsilon Country", and "Counting to Zero" (both from the Center for Thanatology Research in Brooklyn, New York). She has been well-known in both the bereaved-parent and the well-spouse communities for her books and other writings. Some of the activities that keep her sane are classical piano, singing first soprano, Scrabble, thrift-shopping, and spending quality and quantity time with her four living children, two grandchildren, three cats, and Jon.

www.ingramcontent.com/pod-product-compliance
Lightning Source LLC
Chambersburg PA
CBHW072149020426

42334CB00018B/1934